Adriana Manuel

Implementation of Automated Urinary Sediment Analyzer UF1000

Adriana Manuel

Implementation of Automated Urinary Sediment Analyzer UF1000

UF 1000, Sysmex by Roche

Imprint

Any brand names and product names mentioned in this book are subject to trademark, brand or patent protection and are trademarks or registered trademarks of their respective holders. The use of brand names, product names, common names, trade names, product descriptions etc. even without a particular marking in this work is in no way to be construed to mean that such names may be regarded as unrestricted in respect of trademark and brand protection legislation and could thus be used by anyone.

Cover image: www.ingimage.com

This book is a translation from the original published under ISBN 978-620-0-03467-0.

Publisher:
Sciencia Scripts
is a trademark of
Dodo Books Indian Ocean Ltd. and OmniScriptum S.R.L publishing group

120 High Road, East Finchley, London, N2 9ED, United Kingdom
Str. Armeneasca 28/1, office 1, Chisinau MD-2012, Republic of Moldova, Europe
Printed at: see last page
ISBN: 978-620-7-11952-3

Table of Contents :

Chapter 1

IMPLEMENTATION OF AUTOMATED URINARY SEDIMENT ANALYZER UF 1000i (Roche)

Manuel, A (1); Silva, V (2); Prátici, C (3); Fontana, V (3).

1- Biochemist in charge of the Uroanalysis Section. Central Laboratory. Hospital Nuestra Señora del Carmen, OSEP, Mendoza. Contact: adrianamanuel@yahoo.com.ar

2-Technical Uroanalysis Section. Central Laboratory. Nuestra Señora del Carmen Hospital, OSEP, Mendoza.

3-   Clinical Biochemistry Residency. Central Laboratory. Nuestra Señora del Carmen Hospital, OSEP, Mendoza.

The Central Laboratory of OSEP's El Carmen Hospital has incorporated an automated platform for processing isolated urine using Roche's UF 1000 Sysmex flow cytometry technology, which allows us to quantify with high quality and efficiency the formal elements present in these biological samples.

In the Uroanalysis section of the laboratory, the formal elements of the isolated urine samples were subjected to a semi-quantitative study based on the Gold standard method of microscopic observation, which involves the subjectivity of the observer.

The final objective of the present work is to make a comparison of both methods: the semiquantitative microscopic method of daily use and the novel quantitative cytometric method in order to define the parameters that allow the homolagation of the two different methodologies.

DEFINITION OF THE COMPLETE URINALYSIS

Urinalysis is the physical, chemical and microscopic evaluation of urine. This analysis consists of several tests to detect and measure various compounds that come out through the urine.

BIOLOGICAL SAMPLE

A clean urine sample is needed, preferably first thing in the morning using the midstream technique. The sample is kept in a sterile plastic collector with screw cap.

In the laboratory the examination includes the following:

PHYSICAL EXAMINATION:

It depends on the subjectivity of the observer.

✓ Appearance: Cloudy, opalescent, limpid

✓ COLOR: amber yellow is the expected color but depending on the diet and medications

administered can be observed various colors such as blue-green, red, yellow.
Pathologies related to the appearance and color of isolated urine:

- Colorless: Very dilute urine. May be related to polyuria or diabetes insipidus.

- Yellow-orange: concentrated urine; dehydration, fever.

- Brownish yellow: Bilirubin, biliverdin; Hepatopathies.

- Milky: Abundant neutrophils

- Fats (lipuria, chyluria) : Bacterial infections

- Cloudy : - Red cells: urinary tract trauma, hemolytic anemias, infections.

    - Leukocytes: Pyelonephritis, urinary tract inflammation.

    - Fecal contamination: rectovesical fistula.

    - Bacteriuria: Urinary tract infection.

- Red: - Hemoglobin, paroxysmal nocturnal hemoglobinuria, gait hemoglobinuria, glucose 6-P dehydrogenase deficiency, clostridial and Plasmodium falciparum infections.

    - Myoglobin Paroxysmal and gait myoglobinuria, trauma, infections.

    - Hematocytes.   menstrual contamination.

- Red purple: porphyrins, porphyrias.

- Black brown : Homogentisic acid, Alcaptonuria, Methemoglobin, Hemoglobin M, drug-acquired methemoglobinemia.

- Blue-green:- Chlorophyll

        - Pseudomonas

        - Mouth deodorants

CHEMICAL EXAMINATION:

A special strip (test strip) is used to look for various substances in the urine sample. The test strip contains small pads of chemicals that change color when they come in contact with the substances to be tested.

The chemical examination comprises the following tests:

- pH,

- proteins

- glucose

- ketones

- blood

- bile pigments

- urobilinogen

- Nitrites

Reference values for chemical examination

- - Nitrites: Negative

- - pH: 4.6 - 8.0 (average: 6.0)

- - Protein: <0.15 g /24 hours

- - Glucose: Negative

- - Ketones: 17 - 42 mg / dl

- - Bile pigments: Negative

- - Urobilinogen: 0.2 - 1.0 mg / dL

- - Density: - Newborns: 1,012

- - Infants: 1,002 -1,006

- - Adults: 1,001 - 1,035

- - Adults with normal fluid intake: 1.016 -1.024

Since the widespread use of urine test strips, the chemical examination of urine has become a simple and fast procedure.

GLUCOSE

In the case of glucose determination the methods are as follows:

Test strips impregnated with the enzyme glucose oxidase (specific for glucose), for this method the interferences are:

> False positives: urine contaminated with hydrogen peroxide or hypochlorites.

> False negatives: ascorbic acid, ketones, aspirin, infections.

CETONAS

For the determination of ketones in urine, test strips and nitroprusside tablets are used. Feasible interferences are:

> False positives: levodopa, highly colored urine.

> False negatives: ascorbic acid, volatile ketone bodies.

BLOOD

In the case of blood determination in urine, test strips are used for the detection of intact erythrocytes, free hemoglobin and myoglobin.

Possible interferences are:

> False negatives: presence of ascorbic acid with pH less than 5, presence of protein nitrites, increased density.

> False positives: oxidizing contaminants such as hypochlorites.

<u>Chemical parameters modified by different conditions and pathologies.</u>

1- ACID URINE pH<- 7.0

- Diet with high meat protein content.

- Ingestion of some fruits

- Medications such as ammonium chloride, methionine, methenamine mandelate and acid phosphates that are used to acidify urine in the treatment of renal lithiasis.

- In pathological states: respiratory acidosis, metabolic acidosis as in diabetic ketosis, uremia, severe diarrhea and starvation.

- Urinary tract infections due to E. coli.

- In potassium deficit.

2- ALKALINE URINE pH > 7.0

- High intake of vegetables or fruits, especially citrus fruits.

- Medications such as sodium bicarbonate, potassium citrate and acetazolamide used for the treatment of renal lithiasis.

- Treatment with sulfonamides.

- In the treatment of salicylate intoxication.

- Urine collected in the postprandial period.

- In respiratory alkalosis and metabolic alkalosis (vomiting)

- In urinary tract infections caused by urea cleavage germs such as Proteus spp, Pseudomonas spp.

- Samples contaminated with bacteria that take time to process and remain at room temperature. Therefore the elevated pH in urine under these conditions is of no value.

3- PROTEINS:

The presence of proteinuria may be the most important indicator of renal impairment. However, after physical activity, fever, stress and exposure to cold, there may be an increase in protein excretion in the urine.

Normally in the healthy kidney only a small amount of low molecular weight proteins is excreted. This is because the structure of the glomerular membrane does not allow the passage of high molecular weight proteins.

Proteinurias can be classified according to their etiology and the mechanism involved: ❖ Functional not associated with kidney disease: - Excessive exercise.

- Pregnancy

- Orthostatic proteinuria

❖ Organic associated with systemic disease or renal pathology :

- *Pre-renal:* fever, renal hypoxia, hypertension, myxedema, Bence Jones protein.

- *Renal:* glomerulonephritis, nephrotic syndrome and parenchymal lesions.

- *Post-renal:* infection of the pelvis and ureters, cystitis, urethritis or prostatitis.

The type of kidney disease can be predicted by the amount and size of proteins present:
- Minimal proteinuria: less than 0.5 g / 24 hs.

- polycystic kidneys

- chronic pyelonephritis

- chronic inactive glomerulonephritis

- benign orthostatic proteinuria
- Moderate proteinuria: between 0.5 - 3.5 g /24 hs.

- nephrosclerosis

- tubular interstitial disease

- pre-enclampsia

- multiple myeloma

- diabetic nephropathy

- pyelonephritis with hypertension

- Severe proteinuria: greater than 3.5 g / 24 hs.

- glomerulonephritis

- lupus nephritis

- amyloid disease

- lipoid nephrosis

- intercapillary glomerulosclerosis

GLUCOSE

Glucose appears in urine when the blood glucose level exceeds 180 mg/dl. When this happens the renal tubules cannot reabsorb all the filtered glucose and glycosuria occurs.

The most important conditions associated with glycosuria are the following:

1- No hyperglycemia

- Pregnancy

- Renal disease

- Congenital errors

- Contact with nephrotoxic substances (carbon monoxide, mercury)

- Vessel with urine samples contaminated with glucose (leftover sweets, honey)

2- With hyperglycemia

- Diabetes mellitus

- Dietary glycosuria

- Tumors

- Endocrine diseases

- Cushing's syndrome

- Hyperthyroidism

- Pheochromocytoma

CETONAS

They appear in the urine as part of incomplete fatty acid metabolism. In an individual with a normal diet the average value is 20 mg/dl.

Ketonuria is frequently observed in diabetes mellitus.

Clinical significance:

1- Non-diabetic

- Acute febrile and toxic states (with vomiting and diarrhea) in children and infants.

- Vomiting of pregnancy.

- Cachexia.

- Alcoholism.

- Post anesthesia.

- Low carbohydrate diets.

- Prolonged fasting.

2- Diabetic

- Infections in children and young adults.

- Diabetic ketoacidosis

BLOOD

The presence of intact erythrocytes in the urine is called hematuria. Hematuria is also considered when in very alkaline or very low density urine there is lysis of erythrocytes with the release of hemoglobin.

Presence of hematuria:

- Urinary tract pathologies and traumas

- Anticoagulated patients

- Renal lithiasis

- Consumption of some drugs

- hemorrhagic diseases such as hemolytic anemia

- Infections

- Athletes

NITRITES

Many bacteria produce the enzyme reductase, which reduces urinary nitrates to nitrites. This reaction gives color in the reactive area of the strip indicating the presence of bacteria in the urine.

The sensitivity of the test compared to that of a urine culture is only 50%. The test strips are used as a selective test that allows detection of bacteriuria even in cases where it is not clinically suspected.

A positive dipstick result may be an indication for urine culture.

A negative result should not be interpreted as indicating the absence of urinary tract infection, since there are bacteria that do not form nitrites.

BILIRRUBIN

Under normal conditions conjugated bilirubin is not present in urine. It appears in the urine due to extrahepatic biliary tract obstruction (stones in the common bile duct, carcinoma in the head of the pancreas) or intrahepatic (hepatitis, active cirrhosis).

The most sensitive methods for bilirubin detection are Ictotest tablets and test strips.

UROBILINOGENENE

It is produced by the metabolism of intestinal bacteria on conjugated bilirubin. Although the detection of urobilinogen in urine is not part of the routine analysis of whole urine, the use of test strips is useful to know the state of liver function.

Urobilinogen is increased in hemolytic anemias and hepatopathies (hepatitis, cirrhosis).

Interferences: false positives: presence of indole, porphobilinogen.

False negatives: presence of nitrites, formaldehyde.

Chapter 2

PHYSICAL-CHEMICAL STUDY IN LABORATORY

The Uroanalysis section of the Central Laboratory uses an automated platform for the physical and chemical study of urine.

URISYS 2400- ROCHE VS.MANUAL PROCEDURE

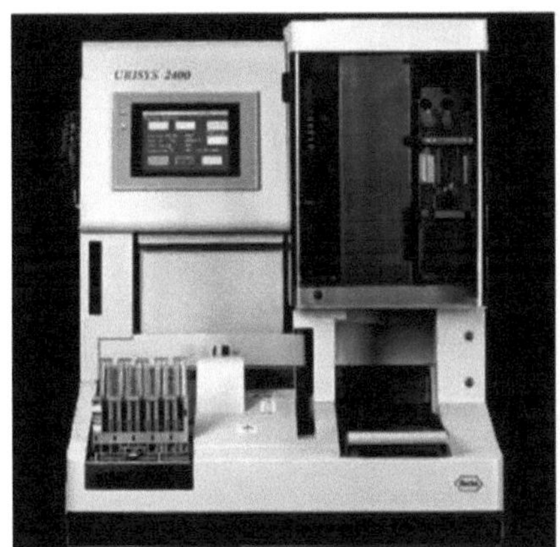
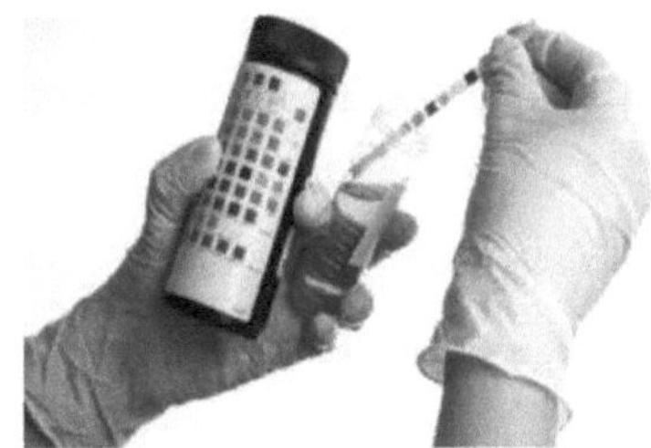

URYSIS 2400 PLATFORM:

Fully automatic system for physicochemical analysis of urine that provides exceptional convenience and efficiency for high and medium volume laboratories.

The Uroanalysis section of the Central Laboratory of Hospital El Carmen receives an average of 300 urines daily for complete urine analysis.

The incorporation of the automated platform allows a much more fluid daily work development, reducing analytical errors to a minimum, thus allowing an accurate report on the alterations or normality of the isolated urine samples that will allow the medical professional a reliable approach to the patient's diagnosis.

TECHNICAL SPECIFICATIONS :

* Innovative, ready-to-use cassette with 400 strips for fast loading in case of high volume work.

* Easy and simple sample loading using standard Roche Diagnostics racks.

* It determines 9 parameters plus color, measures density by refractometry and determines the clarity of the sample.

* Minimum sample volume 1.5 ml.

* Positive sample identification with integrated bar code reader.

* Connectivity through ASTM protocols to the laboratory's central computer system, this allows a direct connection with the laboratory's computer management system; the data obtained in the automated equipment is transmitted avoiding human errors due to manual loading.

Chapter 3

MICROSCOPIC STUDY IN THE LABORATORY

The urine sample is examined under a microscope for:

- Check for cells, urinary crystals, urinary casts, mucus, and other substances.

- Identifying any bacteria or other germs

Urinary sediment: reference values

- **Leukocytes: 0 - 5 / 40 x field**

- **Erythrocytes: 0 - 2 / 40 x field**

- **Epithelial cells: Variable number**

- **Cylinders: Up to 2 hyaline / 10 x field.**

- **Crystals: Variable quantity and characterization depending on condition** acid-alkaline of the sample.

URINE SEDIMENT

Clinical significance:

It is a very useful practice despite its extreme simplicity and low complexity.

Its maximum use will depend on the relationship that the physician makes with the result obtained and the patient's clinical condition.

The urinary sediment is composed of elements of different origins. They can be metabolic products of the kidney such as crystals, cells derived from the bloodstream and urinary tract, cells from other organs of the body, elements originating in the kidney such as casts, and other elements that are not of human origin and appear as contaminants (bacteria and yeasts).

CELLS PRESENT IN URINARY SEDIMENT:

Cells such as erythrocytes or red blood cells, leukocytes or white blood cells and epithelial cells from different parts of the urinary tract, from the tubules to the urethra and also from the vagina or vulva, may be present in the urine as contaminants.

1- Erythrocytes or red blood cells:

The elimination of 0 to 1 or 2 erythrocytes per 40 x field is considered normal. As the erythrocyte membrane is permeable to various solutes in the urine, changes in the shape and size of the erythrocytes depend on the osmotic gradient of the urine, so the erythrocytes appear swollen, crenated or normal in size.

Clinical significance: An increase in the number of red blood cells in the urine (hematuria) indicates lower urinary tract disease or kidney disease.

Hematuria   - All forms of glomerulonephritis

- Renal involvement of systemic diseases

- Benign and malignant tumors of the kidney and urinary tract

- Trauma

- Malformations

- Thrombosis of renal vessels

- Primary infection

- Tuberculosis

- Diabetic nephropathy

- Pyelonephritis

- Hereditary kidney diseases

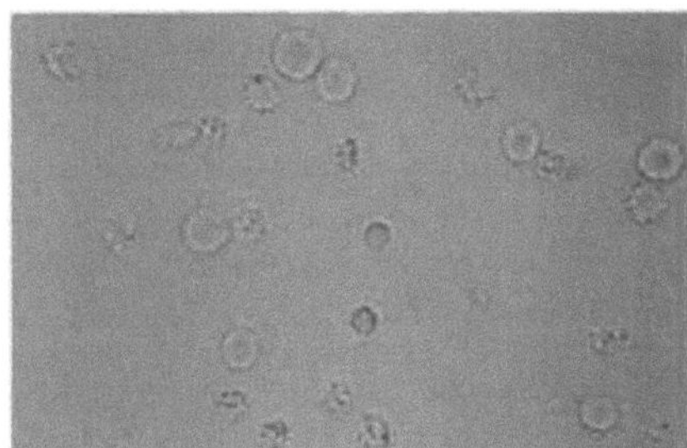

2- White blood cells:

Under abnormal conditions polymorphonuclears are the most frequently found white blood cells in the urinary sediment.

They appear as granulocytes and are characteristic of inflammatory processes of the kidney and urinary tract. It is less common to find lymphocytes, monocytes or eosinophils.

In a normal sediment, 0 to 5 leukocytes per 40 x field are removed.

Clinical significance: an increase in the number of white blood cells in the urine (leukocyturia) represents the fundamental symptom of acute or chronic pyelonephritis, as well as inflammatory diseases of the descending urinary tract such as urethritis, prostatitis, cystitis, pyelitis and tuberculosis.

Leukocyturia:
- Pyelonephritis

- All inflammatory diseases of the descending urinary tract. -

- Glomerulonephritis

- Transplant rejection

- Systemic diseases with renal involvement

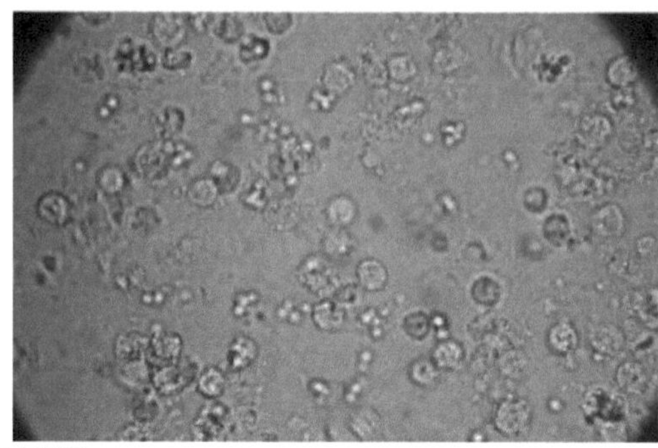

URINE SEDIMENT

<u>**Clinical significance:**</u>

It is a very useful practice despite its extreme simplicity and low complexity.

Its maximum use will depend on the relationship that the physician makes with the result obtained and the patient's clinical condition.

The urinary sediment is composed of elements of different origins. They can be metabolic products of the kidney such as crystals, cells derived from the bloodstream and urinary tract, cells from other organs of the body, elements originating in the kidney such as casts, and other elements that are not of human origin and appear as contaminants (bacteria and yeasts).

<u>**CELLS PRESENT IN URINARY SEDIMENT:**</u>

Cells such as erythrocytes or red blood cells, leukocytes or white blood cells and epithelial cells from different parts of the urinary tract, from the tubules to the urethra and also from the vagina or vulva, may be present in the urine as contaminants.

1- Erythrocytes or red blood cells:

The elimination of 0 to 1 or 2 erythrocytes per 40 x field is considered normal. As the erythrocyte membrane is permeable to various solutes in the urine, changes in the shape and size of the erythrocytes depend on the osmotic gradient of the urine, so the erythrocytes appear swollen, crenated or normal in size.

Clinical significance: An increase in the number of red blood cells in the urine (hematuria) indicates lower urinary tract disease or kidney disease.

Hematuria   - All forms of glomerulonephritis

URINE SEDIMENT

<u>**Clinical significance:**</u>

It is a very useful practice despite its extreme simplicity and low complexity.

Its maximum use will depend on the relationship that the physician makes with the result obtained and the patient's clinical condition.

The urinary sediment is composed of elements of different origins. They can be metabolic products of the kidney such as crystals, cells derived from the bloodstream and urinary tract, cells from other organs of the body, elements originating in the kidney such as casts, and other elements that are not of human origin and appear as contaminants (bacteria and yeasts).

CELLS PRESENT IN URINARY SEDIMENT:

Cells such as erythrocytes or red blood cells, leukocytes or white blood cells and epithelial cells from different parts of the urinary tract, from the tubules to the urethra and also from the vagina or vulva, may be present in the urine as contaminants.

1- Erythrocytes or red blood cells:

The elimination of 0 to 1 or 2 erythrocytes per 40 x field is considered normal. As the erythrocyte membrane is permeable to various solutes in the urine, changes in the shape and size of the erythrocytes depend on the osmotic gradient of the urine, so the erythrocytes appear swollen, crenated or normal in size.

Clinical significance: An increase in the number of red blood cells in the urine (hematuria) indicates lower urinary tract disease or kidney disease.

Hematuria   - All forms of glomerulonephritis

URINE SEDIMENT

Clinical significance:

It is a very useful practice despite its extreme simplicity and low complexity.

Its maximum use will depend on the relationship that the physician makes with the result obtained and the patient's clinical condition.

The urinary sediment is composed of elements of different origins. They can be metabolic products of the kidney such as crystals, cells derived from the bloodstream and urinary tract, cells from other organs of the body, elements originating in the kidney such as casts, and other elements that are not of human origin and appear as contaminants (bacteria and yeasts).

CELLS PRESENT IN URINARY SEDIMENT:

Cells such as erythrocytes or red blood cells, leukocytes or white blood cells and epithelial cells from different parts of the urinary tract, from the tubules to the urethra and also from the vagina or vulva, may be present in the urine as contaminants.

1- Erythrocytes or red blood cells:

The elimination of 0 to 1 or 2 erythrocytes per 40 x field is considered normal. As the erythrocyte membrane is permeable to various solutes in the urine, changes in the shape and size of the erythrocytes depend on the osmotic gradient of the urine, so the erythrocytes appear swollen, crenated or normal in size.

Clinical significance: An increase in the number of red blood cells in the urine (hematuria) indicates lower urinary tract disease or kidney disease.

Hematuria   - All forms of glomerulonephritis

3- Squamous epithelial cells

They originate in the vagina and urethra of both men and women. They may be present in small or large numbers or absent. They are large cells of somewhat irregular appearance with small round nuclei.

Clinical significance: their presence has no pathological value; however, in the presence of squamous cell carcinoma, these cells are affected and undergo changes.

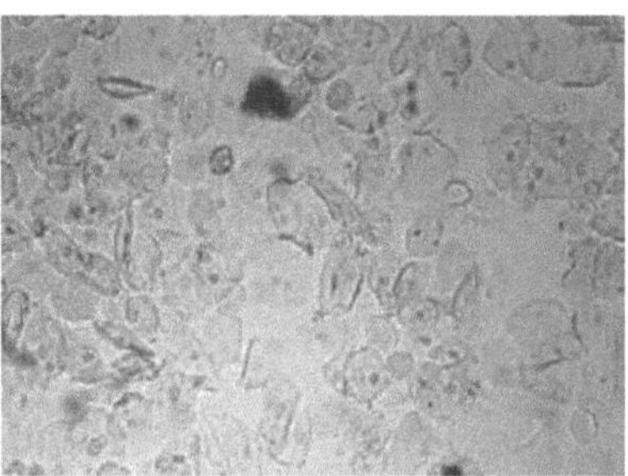

4- Transitional epithelial cells:

They originate from the renal pelvis, ureter and bladder to the urethra. They differ from squamous ones because they are polyhedral to spherical.

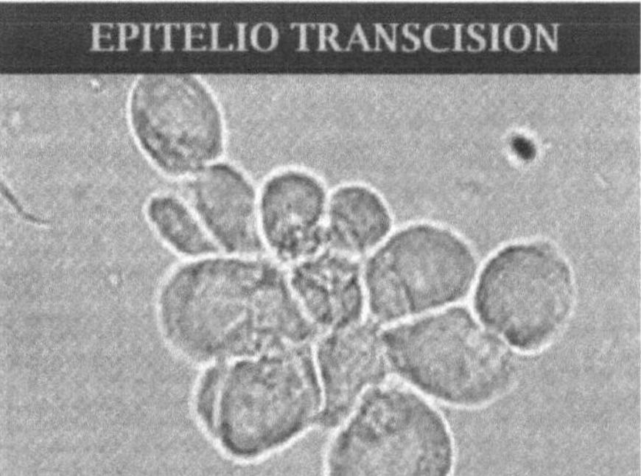

Clinical significance: its presence in large quantities may indicate inflammation of the urinary tract.

5- Epithelial cells of the renal tubule:

They originate from the lining epithelium of the renal tubules. They are difficult to differentiate from transitional leukocytes. They are somewhat larger than leukocytes, have some granulation and their nucleus is not always recognizable.

Clinical significance: They are the most important of all cells from the clinical point of view of the urinary sediment. Their presence in large numbers suggests tubular damage that may occur in diseases such as pyelonephritis, acute tubular necrosis and salicylate intoxication.

They appear in the urinary sediment in patients with viral diseases in general, especially in cytomegalovirus, measles and viral hepatitis, also in toxic lesions (heavy metals) and transplant rejection reactions.

CYLINDER

Cylinder formation occurs in the distal and collecting tubules when the acidification and concentration of urine reaches its maximum extent.

They originate from the thickening or precipitation of proteins and are longitudinal structures that correspond

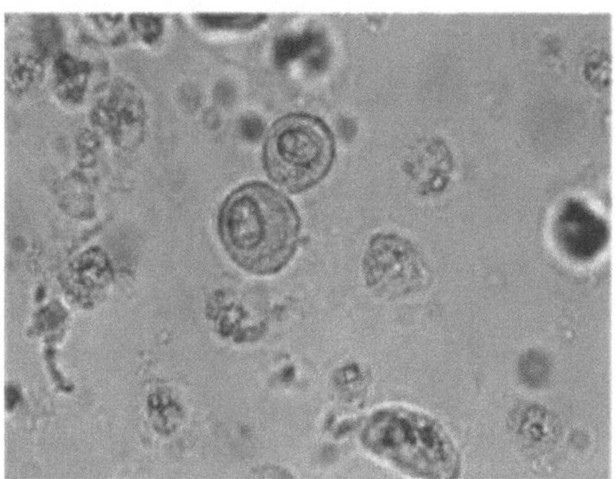

to the lumen of the tubules.

Just as concentrated urine favors the formation of cylinders, dilute urine favors the formation of cylinders.

tend to dissolve. There are different types of cylinders:

1- **Haline cylinders:** These are homogeneous, transparent, colorless, low-refringent structures. In many

hyaline cylinders are observed different types of inclusions that are trapped within them, they can be fine granules, nuclei, cell walls and blood cells. If the hyaline matrix predominates it is considered a hyaline cylinder with inclusions.

Clinical Significance: They are found in the urine of both healthy people and patients with kidney disease. They are also found in the urine of patients receiving certain therapeutic compounds and chemicals that, although not related to kidney disease, affect the kidney in some way.

They are found in large quantities in the sediment of healthy people after great psychic and physical efforts, and also increase with the intake of diuretics such as furosemide and ethacrynic acid.

They can be observed even in the mildest renal disease. They are not associated with any particular disease.

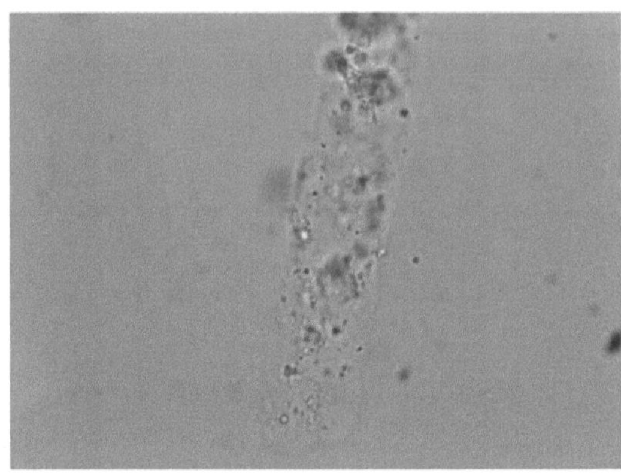

2- **Granular cylinders**: They have morphological characteristics similar to hyaline cylinders, they are usually wider.

and larger than the latter. They have a somewhat higher refractive index than hyaline ones, so they are easier to visualize.

<u>**Clinical significance:**</u> they are present in both normal and non-normal sediments.

They are found in large amounts after physical exertion in healthy people and are otherwise frequently associated with acute and chronic kidney disease, especially in glomerulonephritis and more rarely in pyelonephritis.

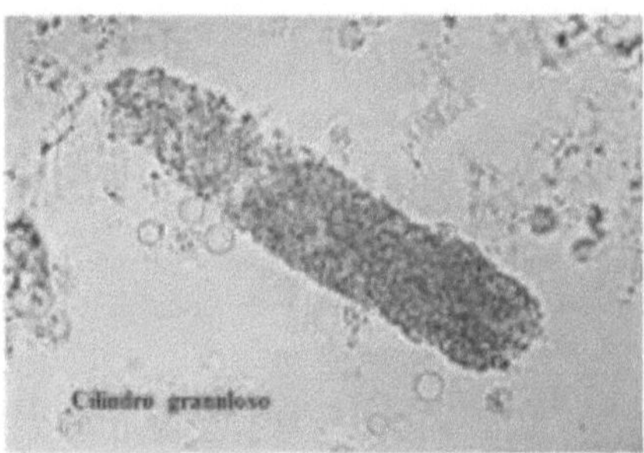

3- **Waxy cylinders:** They are easily recognized in a common field of view because they have an index of

refraction greater than that of all cylinders in general, for its broken or abruptly terminated tips, as well as for its characteristic notches or fine indentations on the edges, which are perpendicular to the longitudinal axis of the same. They are slightly yellow in color.

<u>**Clinical significance**</u>: its presence in the urine always indicates severe chronic kidney disease.

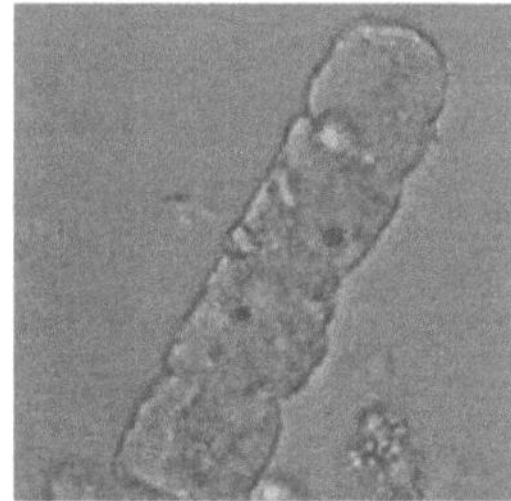

4- **Erythrocyte cylinders :**They are composed of more or less dense erythrocytes that adhere to a substance.

> hyaline. Their color varies from yellowish red to brownish, although they can be lighter and even colorless.

As the degeneration of the erythrocyte casts progresses, the boundaries disappear and the so-called yellowish-red hematic casts are formed. .

Clinical significance: they are indicators of glomerular injury. They are often found in diseases like glomerulonephritis, lupus erythematosus and more rarely in bacterial endocarditis .

Erythrocyte or hematic casts always indicate hematuria of renal origin.

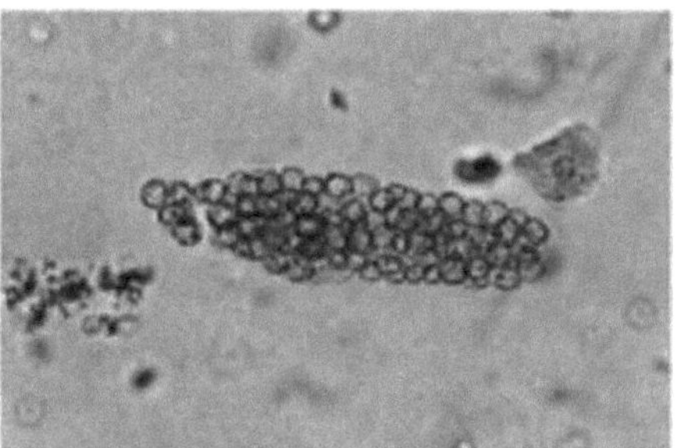

5- **Leukocyte cylinders:** They are formed by a few leukocytes or by many of these cells agglomerated,

> which adhere to the cilium through a hyaline matrix Leukocyte casts occur in the presence of intense intrarenal exudation of leukocytes and elimination of proteins through the tubules.

Most of the leukocytes in the casts are polymorphonuclear neutrophils.

Clinical significance: they are not found in normal sediment. Most of the times they are associated with renal infections. They are observed in 80% of pyelonephritis cases, they are also observed in glomerulonephritis.

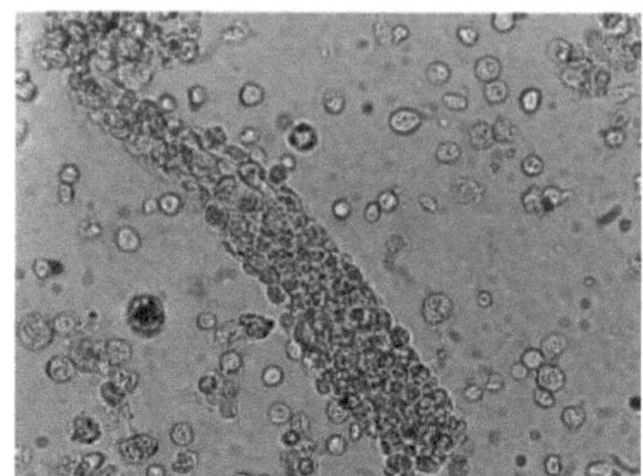

LEUKOCYTE CYLINDER

CRYSTALS

They are normally present in all urine, the most important thing is to know how to differentiate normal urine crystals from those associated with some pathology.

When urine is supersaturated with a particular crystalline compound or when the solubility properties of the urine are altered, the formation of crystalline compounds occurs.

Amorphous crystals of urates, uric acid and calcium oxalates are observed in acid urine, while phosphate crystals are always found in alkaline urine.

Crystals can take different forms depending on the chemical compound and the pH of the urine.

✓ **Uric acid crystals**: They exist in various forms, rhomboidal squares, grinding stone, rosettes, weights, barrels and canes.

Their color varies from brownish red to colorless.

Clinical significance: Its presence in urine does not necessarily indicate a pathological condition.

They are present in the urine in diseases such as gout, leukemia, increased purine metabolism,

acute febrile illness and chronic nephritis.

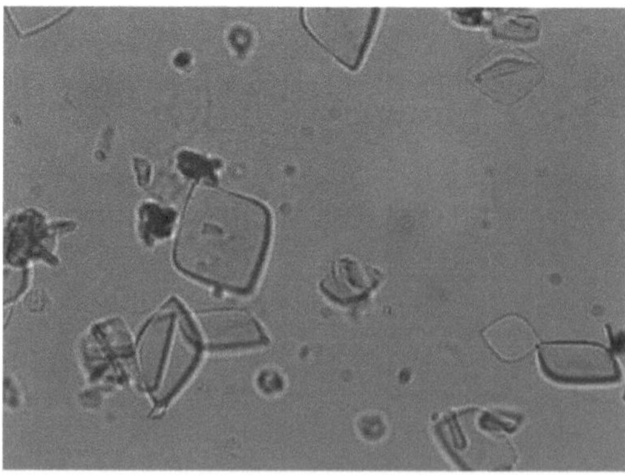

✓ **Amorphous urates:**
Uric acid salts are found in acidic or neutral urine, in non-crystalline, amorphous form. Sodium, potassium,

magnesium and calcium urates can be found. They have a granular appearance and may be pink or reddish-yellow in color. This precipitate is known as brick dust.

<u>Clinical significance:</u> they are frequent in concentrated urine as in the case of fever and also in gout, but lack diagnostic significance.

The precipitate is amber in color and dissolves with heat.

✓ **Calcium oxalates: They** are usually found in acidic urine, although they can also form in urine with a slightly alkaline to neutral pH. They are colorless, octahedral or envelope-shaped, simulating small squares crossed by intersecting diagonal lines. Other times they appear as oval spheres or biconcave discs shaped like gymnastic weights.

<u>Clinical significance:</u> All forms can be found in a normal sediment, depending on the diet. Their number increases when the diet is rich in oxalic acid (tomatoes, oranges, asparagus, and apples). These crystals are related to the formation of kidney stones and have been seen in large numbers in patients with pathologies such as diabetes mellitus, nervous system diseases, liver disease and chronic kidney disease.

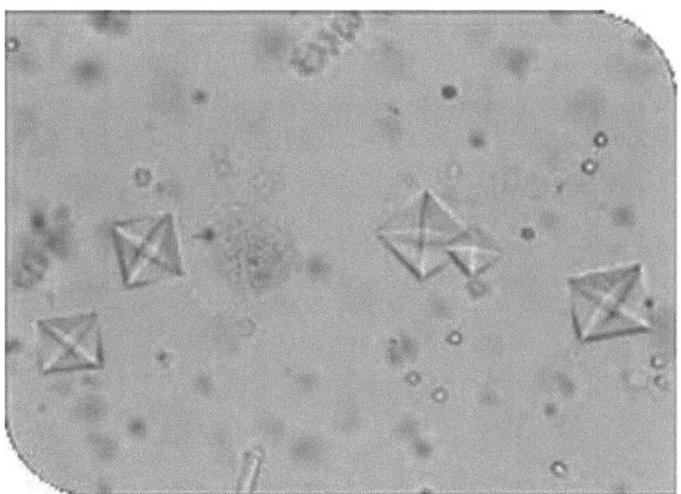

✓**Hippuric acid crystals**: infrequently observed, may form in slightly alkaline urine. or neutral but are always found in acidic urines.

They are colorless or have a pale yellow color. They are observed as elongated prisms or plates, may be so thin that they look like needles and are often grouped together.

<u>Clinical significance:</u> They are usually absent, but have been found in large numbers in patients with acute febrile state and in liver disease.

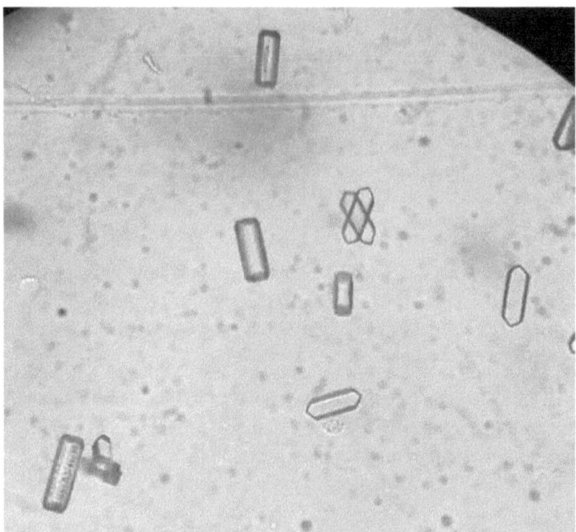

Amorphous phosphate crystals: They appear in neutral and alkaline urine as fine, colorless granules that tend to occur in clusters. They have no clinical significance.

The sediment is whitish and dissolves with the addition of a drop of acetic acid.

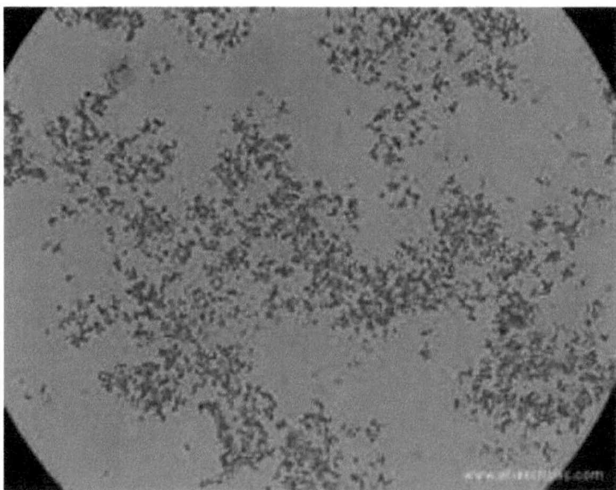

Triple phosphate crystals: Also called ammonium magnesium phosphates, they occur in neutral and alkaline urines. They occur as colorless 3- to 6-sided prisms that often have oblique ends. Sometimes they can precipitate forming feathery or fern-like crystals.

Clinical significance: they appear in pathological processes such as chronic pyelitis, chronic cystitis, prostate hypertrophy and in cases of bladder retention of urine. They can form urinary calculi.

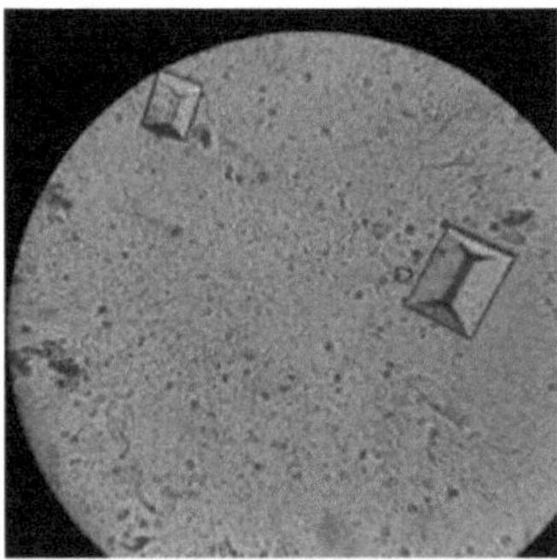

✓ **Triple calcium phosphate crystals**: Also known as dicalcium phosphates, they appear in alkaline urine.

They can be found in the form of amorphous granules and also in crystalline forms. The most common form is that of a large irregular plate resembling a sheet of ice.

Clinical significance: although they have no clinical **significance**, they are associated with patients with cystitis with urine retention and with the formation of kidney stones.

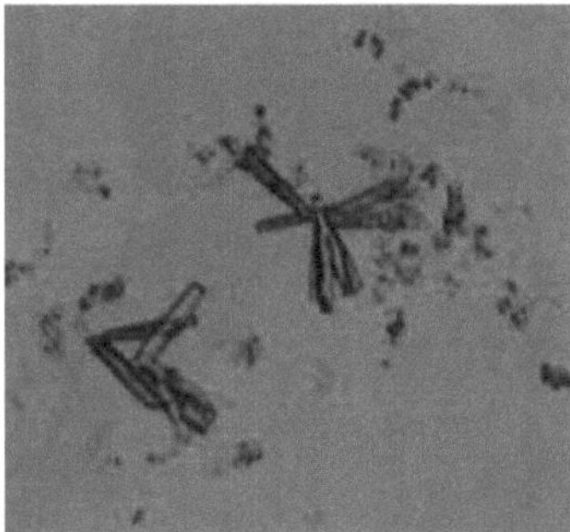

✓ **Ammonium urate crystals**: These are the only urate crystals found in alkaline urine. They are spherical brownish-yellow bodies with long irregular spicules or without them.

Clinical significance: they have no clinical **significance**, but constitute an abnormality only if they are found in freshly emitted urine. They appear in the formation of ammonium in bladder urine.

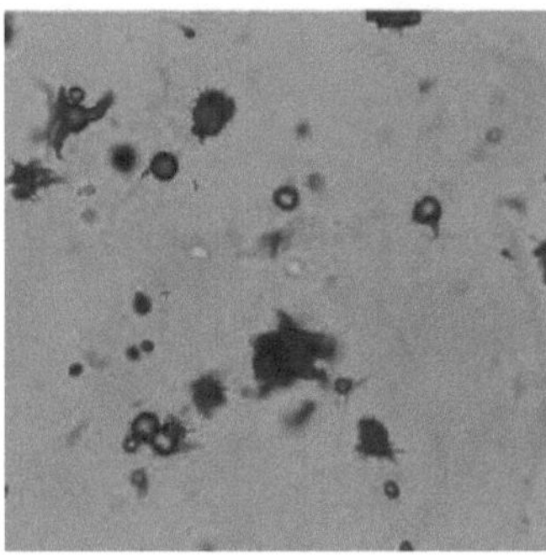

✓ **Leucine crystals: They are** found in acid urine in the form of spheres with concentric striations.

They are highly refractive and appear as yellowish or brownish bodies. They appear in urine in association with tyrosine crystals.

Clinical significance: they respond to the same conditions as tyrosine tyrosines.

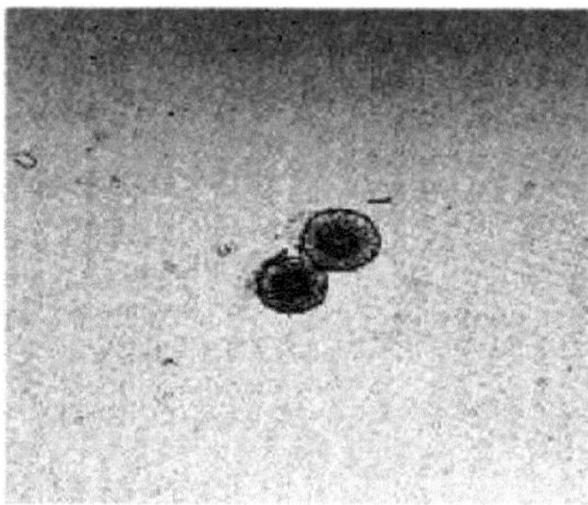

✓ **Cystine crystals**: They are found in urine with acid pH and are observed as thin, colorless, hexagonal sheets.

Clinical significance: they are most often observed in the urine of patients suffering from various types of inherited metabolic disorders.

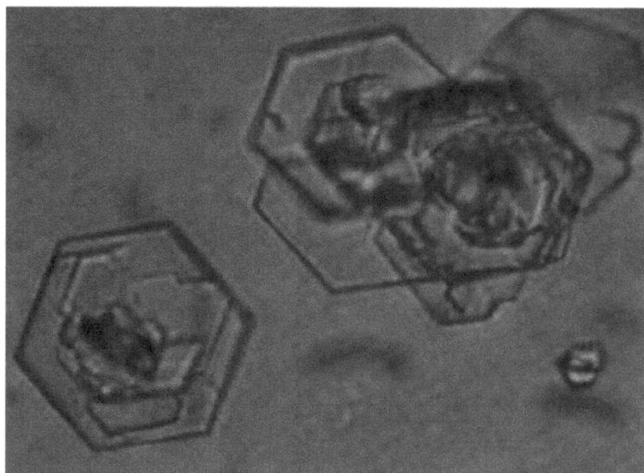

✓ **Tyrosine crystals**: They are very rare and are only observed in acid urine. Their color varies from colorless to brownish yellow. Their shape is that of very fine and refringent needles, appearing in groups or clusters.

Clinical significance: They are frequently found together with leucine crystals. They are a product of protein metabolism. Clinical significance: They appear in urine of patients with tissue necrosis or degeneration such as acute liver disease, hepatitis, cirrhosis, leukemia and typhoid fever.

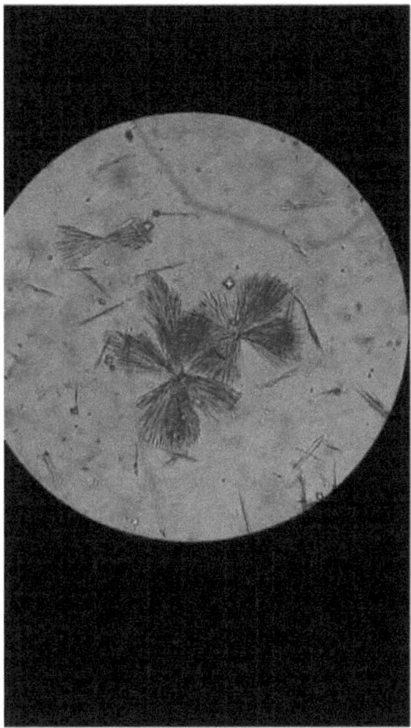

✓ **Cholesterol crystals**: Found in acidic or neutral urine, they appear as flat, transparent sheets with notched angles. They are often found forming a film on the surface of the urine instead of being found in the sediment.

Clinical significance: they are not common in urine and whenever they are present they are always related to some pathology.

They are found in renal diseases such as nephrotic syndrome and predominate in chyluria, which occurs as a consequence of obstruction of the lymphatic flow in the abdomen.

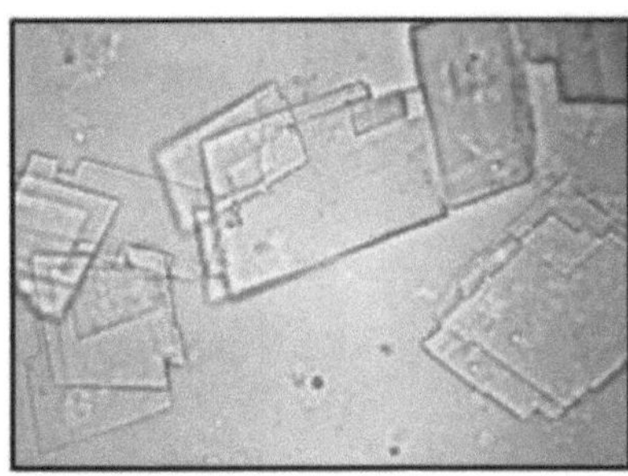

BACTERIA

There are no bacteria in the kidneys or bladder. Although urine is free of bacteria, it can be contaminated with bacteria present in the urethra or vagina.

Clinical significance: when a urine sample is collected in sterile form and contains a large number of bacteria and is accompanied by many leukocytes, a urinary tract infection is very likely to be found.

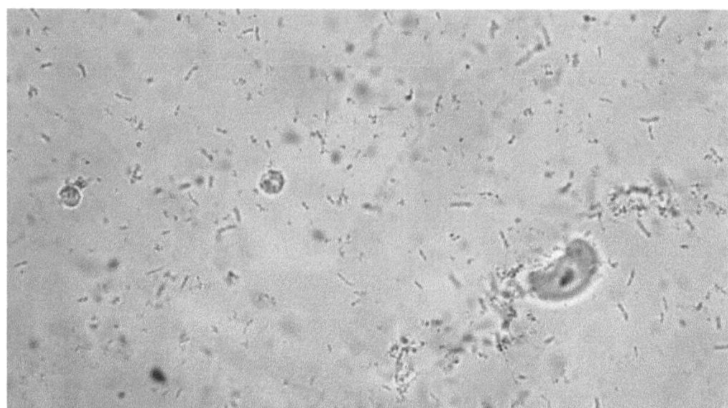

FUNGI

They are oval-shaped colorless structures. Sometimes they can be confused with erythrocytes, but they are somewhat smaller than erythrocytes and often have tubular or filamentous evaginations (hyphae).

Clinical significance: They are commonly found in patients with metabolic diseases (diabetes mellitus). They are commonly found in urine sediments belonging to pregnant women.

If found in sediment with vaginal contamination, it is suggested not to report.

They are recognized as having pathological value in patients with low defenses, in these cases it is Candida albicans that plays a fundamental role.

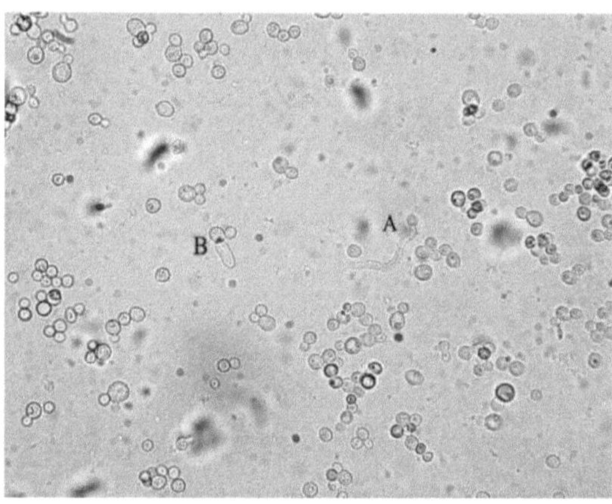

MUCUS

They are long, thin, wavy, irregular, ribbed filaments of variable length.

Epithelial cells, leukocytes, erythrocytes and even crystals often hang from these mucous filaments.

Clinical significance: they normally exist in the urine in small amounts, but may be very abundant in case of inflammation or irritation of the urinary tract.

In urine with vaginal contamination it is usually found to be increased.

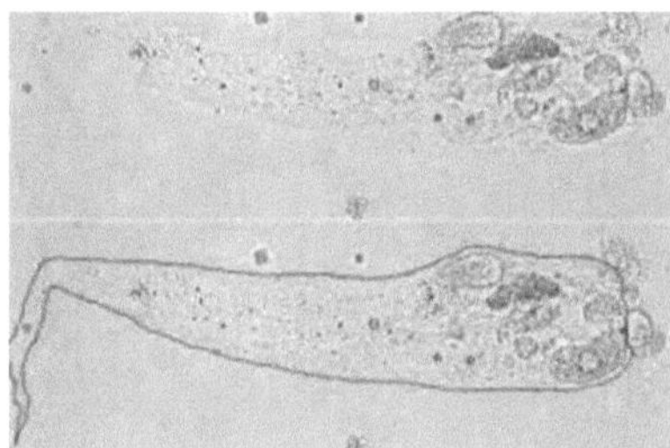

SPERMATOZOA:

They are normally found in male urine because they anatomically share the ureter with the seminal duct. It is suggested to report their presence regardless of the patient's age.

In female urine they are found as contamination and should not be reported.

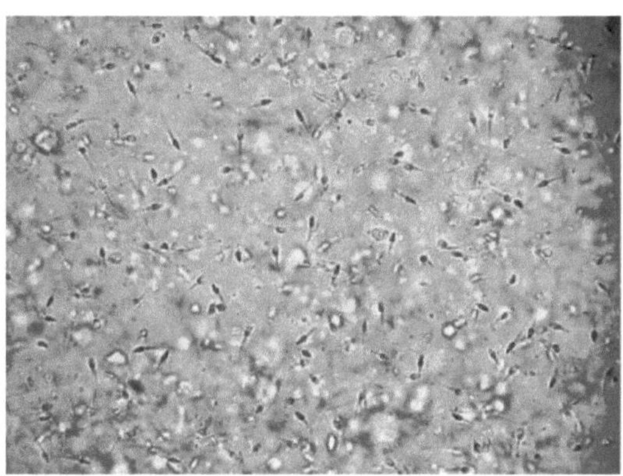

CLINICAL UTILITY OF COMPLETE URINE:

-   Screening in the general population.

- Auxiliary evaluation of renal function. Through the complete urine, pathological alterations of the kidney and urinary tract are revealed.

-   Monitoring therapy of urinary tract disorders.
-   As part of a routine medical examination for the early signs of disease

- If the patient has signs of diabetes or renal disease, or to monitor whether the patient is receiving treatment for such

conditions

-   To check for the presence of blood in the urine

-   To diagnose urinary tract infections

Certain medications change the color of the urine, but this is not a sign of disease.

Medications that can change the color of urine include:

Chloroquine

Iron supplements

Levodopa

Nitrofurantoin

Phenazopyridine

Phenothiazine

Phenytoin

Riboflavin

Triamterene

Abnormal results may mean that pathologies may be present such as:

Urinary tract infection

Kidney stones

Poorly controlled diabetes

Gallbladder or kidney cancer

IMPLEMENTATION OF UF 1000 PLATFORM IN LABORATORY

INTRODUCTION

We defined the concentration thresholds of urinary parameters such as red blood cells, white blood cells, crystals, casts, bacteria, to implement the automated urinary sediment analyzer UF 1000i (Roche). This study ensures the correct functioning of the analyzer, offers better technical support and increases the level of professionalism of our work.

OBJECTIVES

Set reference and alarm intervals for the UF 1000i automated urine sediment analyzer.

MATERIALS AND METHODS

A total of 162 urine samples from healthy individuals were processed during August 2014. The results obtained by UF 1000i were compared with those obtained by the reference method of microscopic observation. The EP Evaluator tool was used to perform the statistical analysis, using a qualitative and comparative method.

Chapter 4

RESULTS OBTAINED BY LINEARITY AND COMPARISON OF MICROSCOPIC VS. MICROSCOPIC METHOD. UF 1000

	OBSERVED		NO OBSERVATIONS	
	UF 1000 Part./ul	MICROSCOPY Part/field 45x	UF1000 Part./ul	MICROSCOPY Part/field 45x
EPITHELIAL CELLS	Greater than 32	Regular or abundant quantity	Less than or equal to 31	Absence or Scarcity
LEUCOCYTES	Greater than 29	Greater than 3	Less than or equal to 28	0-3
HEMATÍES	Greater than 24	Greater than 5	Less than or equal to 23	0-5
CRYSTALS	Greater than 4	Presence	Less than or equal to 3	Absence
HYALINE CYLINDERS	Greater than 34	Presence	Less than or equal to 33	Absence
PATHOLOGICAL CYLINDERS	Greater than 34	Presence	Less than or equal to 33	Absence

STATISTICAL RESULTS USING EP EVALUATOR SYSTEM
Qualitative Method Comparison Summary
Epithelial Cells

Qualitative Method Comparison

Ref Method Microscopy Test Method UF1000

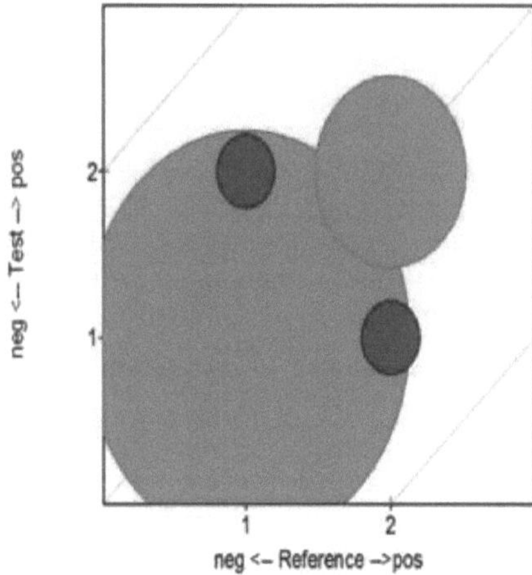

Statistical Analysis
(Comparison of two Laboratory Methods)

Agreement94   .9%   (87.5 to 98.0%)
Positive Agreement86   .7% Positive Agreement86   .7%
Positive Agreement86   .7% Positive Agreement86   .7%
Positive Agreement
Negative Agreement96   .8%.
95% confidence interval calculated by the "Score" method.

McNemar Test for Symmetry:
Test <   Reference2 (2.6%)
Test >   Reference2 (2.6%)
Symmetry test PASSES p > 0,999 (Exact Test)
A value of p<0.05 suggests that one method is consistently "larger".

Cohen's Kappa83   .5% (67.7 to 99.2%)
Kappa is the portion of agreement above what is expected by chance
The rule of thumb is that Kappa > 75% indicates "high" agreement.
We would like to see VERY high agreement (close to 100%).

Statistical Summary

	Negative Reference	Positive Reference	Total	Legend		ReferenceTest
Negative	61	2	63	1 Scarce (E)		Negative
<I> Test						
Positive Test	2	13	15			
Total	63	15	78			

Number excluded or missing:

EP Evaluator®

Experiment Description

	Reference Method	Test Method
Analyst:	Technician	Technician
Date:	06 Aug 2014	06 Aug 2014
Comment		

Epithelial Cells

Qualitative Method Comparison

Ref Method Microscopy Test Method UF1000

Experimental Results

Spec ID	Ref	Test	Spec ID	F	Ref	Test	Spec ID	F	Ref	Test
*0214	E	0,20	*3031			R45	*8344			E0 ,29
*1363	E	3,70	*3032	F	R	29	*8357			E0 ,50
*1462	E	4,50	*3033	F		E47	*8412			E3 ,60
*2017	E	5,10	*3034			E23	*8482			R37 ,40
*2305	E	7,80	*3035			E13	*8489			E21 ,80
*2306	E	5,10	*3036			E21	*8504	F	R9	,30
*2327	E	5,30	*3037			R67	*8513	F	E60	,00
*2329	E	28,80	*3038			R69	*8515			R37 ,70
*2330	E	0,20	*3039			E12	*8520			E6 ,20
*2338	E	0,10	*3040			E18	*8521			E1 ,80
*2362	E	2,30	*3041			R49	*8522			E2 ,40
*2367	E	4,50	*3066		E24	,40	*8523			E9 ,10
*2376	R	82,60	*3081		E0	,40	*8524			E4 ,80
*2377	E	1,10	*3165		E16	,00	*8526			E0 ,50
*2812	E	3,00	*3339		E0	,00	*8527			E15 ,80
*2860	E	0,10	*3341		E7	,10	*8528			E1 ,80
*3021	E	17,60	*3344		E1	,30	*8529			E0 ,80
*3022	E	17,60	*3345		E28	,80	*8530			E4 ,90
*3023	E	21	*3482		E13	,90	*8531			E1 ,60
*3024	E	14	*3741		E14	,80	*8532			E0 ,70
*3025	R	45	*7964		E4	,60	*8533			E2 ,10
*3026	E	21	*8168		E5	,30	*8534			E12 ,20
*3027	E	10,6	*8305		E0	,70	*8857			E9 ,50
*3028	E	24	*8321		R37	,10	*8967			R39 ,20
*3029	R	109	*8325		R39	,70	*8987			E9 ,60
*3030	R	39	*8336		E7	,20	*9985			E0 ,80

X:Excluded F:No Agreement

Cyl. Hyaline

Prepared for: Central Laboratory -- Hospital El Carmen
By: Analytical Quality -- Productos Roche S.A.Q.e I

Qualitative Method Comparison

Ref Method Microscopy   **Test Method UF1000**

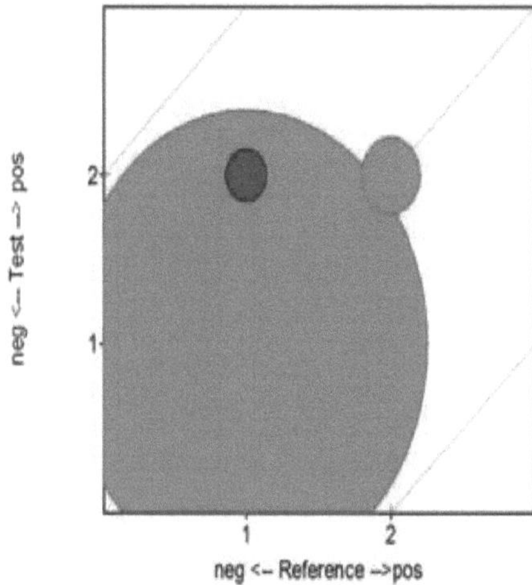

Statistical Analysis
(Comparison of two Laboratory Methods)

Agreement98   .7% (93.1 to 99.8%)
Positive Agreement100 .0%
Negative Agreement98   .7% Negative Agreement98
                       95% confidence interval
calculated by the "Score" method.
McNemar Test for Symmetry:
Test <   Reference0 (0.0%)
Test >   Reference1 (1.3%)
Symmetry test PASSES p > 0.999 (Exact Test)
A value of p<0.05 suggests that one method is consistently
"larger".

Cohen's Kappa79   .4% (39.2 to 119.5%)
Kappa is the portion of agreement above what is expected
by chance The rule of thumb is that Kappa > 75%
indicates "high" agreement. We would like to see VERY
high agreement (close to 100%).

Statistical Summary

	Negative Reference	Positive Reference	Total	Legend Reference	Test
Negative Test	75	-	75	1 Negative (NSO)	Negative
Positive Test	1	2	3	2 Positive (3)	(<=33) Positive (>33)
Total	76	2	78		

Number excluded or missing: 5

Experiment Description

	Reference Method	Test Method
Analyst:	Technician	Technician
Date:	06 Aug 2014	06 Aug 2014
Comment		

Cyl. Hyaline

Prepared for: Central Laboratory -- Hospital El Carmen
By: Analytical Quality -- Productos Roche S.A.Q.e I

Qualitative Method Comparison

Ref Method Microscopy   **Test Method UF1000**

	Experimental Results							
Spec ID	Ref	Test	Spec ID	Ref	Test	Spec ID	Ref	Test
*0214	NSO	0,00	*3741	NSO	0,00	*8534	NSO	0,43
*1363	NSO	0,72	*7964	NSO	19,09	*8857	NSO	2,97
*1462	F NSO	264	*8168	NSO	0,58	*8967	NSO	9,72
*2017	NSO	0,14	*8305	NSO	0,14	*8987	NSO	0,14
*2305	NSO	0,43	*8321	NSO	13,41	*9985	NSO	0,29
*2306	NSO	0,72	*8325	NSO	15,30	3540	X4	65
*2327	NSO	0,00	*8336	NSO	1,25	4769	X4	54
*2329	NSO	0,72	*8344	NSO	0,29	4770	NSO	1,8
*2330	NSO	0,29	*8357	NSO	0,14	4771	NSO	1,9
*2338	NSO	0,00	*8412	NSO	1,02	4772	NSO	24
*2362	NSO	0,43	*8482	NSO	3,64	4773	NSO	6,9
*2367	NSO	0,43	*8489	NSO	3,64	4774	O XNS	6.0
*2376	NSO	0,87	*8504	NSO	1,16	4775	NSO	2,9
*2377	NSO	0,14	*8513	NSO	30,41	4776	NSO	23
*2812	NSO	0,29	*8515	NSO	1,02	4777	NSO	0,98
*2860	NSO	0,14	*8520	NSO	0,43	4778	3	37
*3021	NSO	0,58	*8521	NSO	0,14	4779	NSO	0,87
*3022	NSO	0,58	*8522	NSO	0,14	4780	NSO	0,6
*3023	NSO	1,9	*8523	NSO	0,29	4781	O XNS	0.87
*3024	NSO	12	*8524	NSO	0,29	4782	NSO	1,6
*3066	NSO	7,14	*8526	NSO	0,14	4783	NSO	2,9
*3081	NSO	0,00	*8527	NSO	1,16	4784	O XNS	0.76
*3165	NSO	1,31	*8528	NSO	0,29	4785	NSO	0,5
*3339	NSO	0,29	*8529	NSO	0,00	4786	NSO	1,2
*3341	NSO	0,58	*8530	NSO	0,43	4787	3	34
*3344	NSO	0,29	*8531	NSO	6,26	4788	NSO	0,98
*3345	NSO	0,72	*8532	NSO	0,14	4789	NSO	1,6
*3482	NSO	0,41	*8533	NSO	0,00			

X:Excluded F:No Agreement

Prepared for: Central Laboratory -- Hospital El Carmen
By: Analytical Quality -- Productos Roche S.A.Q.e I

Qualitative Method Comparison

Ref Method Microscopy   **Test Method UF1000**

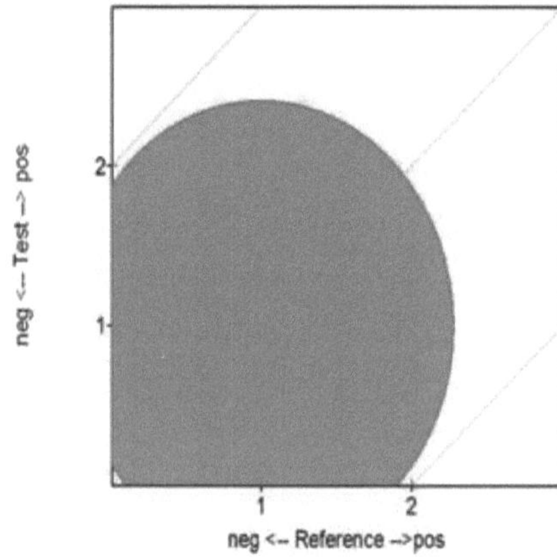

Statistical Analysis
(Comparison of two Laboratory Methods)

Agreement100   .0%   (95.3rd 100.0%)
Positive Agreement   -
Negative Agreement100   .0% 100   .0%
95% confidence interval calculatedby the "Score" method.

McNemar Test for Symmetry:
Test <   Reference0   (0.0%)
Test >   Reference0   (0.0%)

Symmetry test PASSES

Cohen's Kappa100   .0% (100.0 to100   .0%)
Kappa is the portion of agreement above what is expected
by chance The rule of thumb is that Kappa > 75% indicates
"high" agreement. We would like to see VERY high
agreement (close to 100%).

Statistical Summary

	Negative Reference	**Positive Reference**	**Total**	**Legend**	
				Reference	**Test**
Negative Test	78	-	78	1 Negative (NSO)	Negative (<=33) Positive (>33)
Positive Test	-	-	0	2 Positive (1)	
Total	78	0	78		

Number excluded or missing: 2

Experiment Description

	Reference Method	**Test Method**
Analyst:	Technician	Technician
Date:	06 Aug 2014	06 Aug 2014
Comment		

Prepared for: Central Laboratory -- Hospital El Carmen
By: Analytical Quality -- Productos Roche S.A.Q.e I

Qualitative Method Comparison

Ref Method Microscopy Test Method UF1000

Experimental Results

Spec ID	Ref	Test	Spec ID	Ref	Test	Spec ID	Ref	Test
*0214	NSO	0,0	*3075	NSO	2,9	*8515	NSO	0,7
*1363	NSO	0,6	*3076	NSO	1,8	*8520	NSO	0,1
*1462	NSO	0,0	*3077	NSO	1,4	*8521	NSO	0,1
*2017	NSO	0,1	*3078	X NSO	0.5	*8522	NSO	0,1

*2305	NSO	0,4	*3079	NSO	0,78	*8523	NSO	0,3
*2306	NSO	0,7	*3080	NSO	1,45	*8524	NSO	0,3
*2327	NSO	0,1	*3081	NSO	0,0	*8526	NSO	0,1
*2329	NSO	0,7	*3165	NSO	1,2	*8527	NSO	0,9
*2330	NSO	0,3	*3339	NSO	0,0	*8528	NSO	0,3
*2338	NSO	0,0	*3341	NSO	0,6	*8529	NSO	0,0
*2362	NSO	0,4	*3344	NSO	0,3	*8530	NSO	0,4
*2367	NSO	0,4	*3345	NSO	0,0	*8531	NSO	0,2
*2376	NSO	0,9	*3482	NSO	0,0	*8532	NSO	0,0
*2377	NSO	0,0	*3741	NSO	0,0	*8533	NSO	0,0
*2812	NSO	0,1	*7964	NSO	4,4	*8534	NSO	0,1
*2860	NSO	0,1	*8168	NSO	0,1	*8857	NSO	0,0
*3021	NSO	0,4	*8305	NSO	0,1	*8967	NSO	3,4
*3022	NSO	0,4	*8321	NSO	7,6	*8987	NSO	0,1
*3066	NSO	2,6	*8325	NSO	11,4	*9985	NSO	0,3
*3067	NSO	2,6	*8336	NSO	0,3	3082	NSO	0,4
*3068	NSO	1,9	*8344	NSO	0,3	3083	NSO	0,5
*3069	NSO	2,9	*8357	NSO	0,1	3084	NSO	0,80
*3070	X NSO	0.9	*8412	NSO	1,0	3085	NSO	0,5
*3071	NSO	1,8	*8482	NSO	0,7	3086	NSO	0,0
*3072	NSO	0,0	*8489	NSO	1,0	3087	NSO	0,76
*3073	NSO	1,6	*8504	NSO	0,9	3088	NSO	0,76
*3074	NSO	3,9	*8513	NSO	11,3			

X:Excluded F:No Agreement

Qualitative Method Comparison

Ref Method Microscopy   **Test Method UF1000**

Statistical Summary

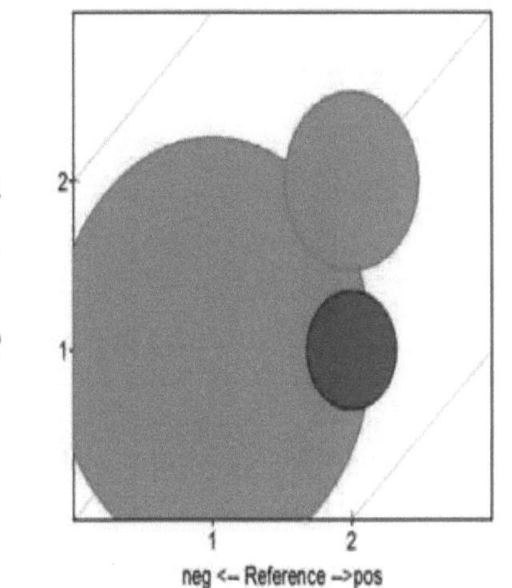

Statistical Analysis
(Comparison of two Laboratory Methods)

Agreement93   .6% (85.9% to   97.2%)
Positive Agreement68   .8%.
Negative Agreement100   .0% 100   .0%
95% confidence interval calculated by the "Score" method.

McNemar Test for Symmetry:
Test <   Reference5   (6.4%)
Test >   Reference0   (0.0%)
Symmetry test PASSES p = 0.063 (Exact Test)
A value of p<0.05 suggests that one method is consistently "larger".

Cohen's Kappa77   .8% (58.9 to 96.6%)
Kappa is the portion of agreement above what is expected by chance The rule of thumb is that Kappa > 75% indicates "high" agreement. We would like to see VERY high agreement (close to 100%).

	Negative Reference	Positive Reference	Total	Legend		
					Reference	Test
Negative Test	62	5	67	1Negative (NSO)	Negative	
Positive Test	-	11	11	2 Positive (ESC.)		(<=3) Positive (>3)
Total	62	16	78			

Number excluded or missing: 5

Experiment Description		
	Reference Method	Test Method
Analyst:	Technician	Technician
Date:	06 Aug 2014	06 Aug 2014
Comment		

Prepared for: Central Laboratory -- Hospital El Carmen
By: Analytical Quality -- Productos Roche S.A.Q.e I

Qualitative Method Comparison

Ref Method Microscopy Test Method UF1000

Experimental Results								
Spec ID	Ref	Test	Spec ID	Ref	Test	Spec ID	Ref	Test
*0214	NSO	0	*3174	NSO	0	*8336	NSO	0
*1363	FESC	1	*3175	NSO	0	*8344	NSO	0
*1462	NSO	0	*3176	NSO	0	*8357	XEOC	2
*2017	NSO	2	*3177	NSO	0	*8412	NSO	0
*2305	NSO	0	*3178	NSO	0	*8482	NSO	0
*2306	NSO	0	*3179	NSO	1	*8489	NSO	0
*2327	NSO	0	*3180	NSO	1	*8504	NSO	0
*2329	NSO	0	*3181	ESC.	5	*8513	ESC.	211
*2330	NSO	0	*3182	NSO	2	*8515	FESC	0
*2338	NSO	1	*3183	ESC.	38	*8520	NSO	0
*2362	XEOC	0	*3184	NSO	1	*8521	NSO	0
*2367	XEOC	0	*3185	NSO	0	*8522	NSO	0
*2376	NSO	0	*3186	NSO	0	*8523	ESC.	442
*2377	XEOC	39	*3187	NSO	0	*8524	NSO	0
*2812	ESC.	895	*3188	ESC.	36	*8526	NSO	0
*2860	NSO	0	*3189	ESC.	45	*8527	NSO	0
*3021	NSO	0	*3190	NSO	0	*8528	FESC	2
*3022	NSO	0	*3339	NSO	0	*8529	NSO	1
*3066	ESC.	772	*3341	NSO	0	*8530	NSO	0
*3067	ESC.	772	*3344	NSO	0	*8531	NSO	0
*3081	NSO	0	*3345	NSO	0	*8532	ESC.	49
*3165	NSO	2	*3482	F ESC.	1	*8533	NSO	0

*3168	NSO	2	*3741	NSO	0	*8534	NSO	2
*3169	NSO	1	*7964	NSO	0	*8857	*FESC*	*3*
*3170	NSO	0	*8168	NSO	0	*8967	NSO	0
*3171	NSO	0	*8305	NSO	0	*8987	NSO	2
*3172 *3173	NSO NSO	1 2	*8321 *8325	ESC. NSO	133 0	*9985	*XEOC*	*89*

X:Excluded F:No Agreement

Prepared for: Central Laboratory -- Hospital El Carmen
By: Analytical Quality -- Productos Roche S.A.Q.e I

Qualitative Method Comparison

Ref Method Microscopy **Test Method**
UF1000

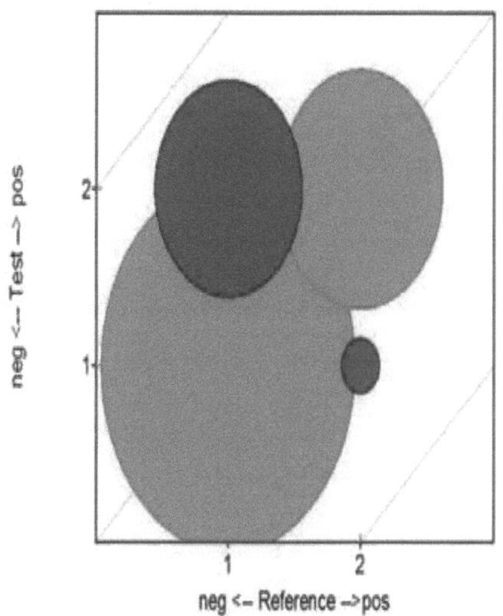

Statistical Analysis
(Comparison of two Laboratory Methods)

Agreement79 .5% (69.2 to 87.0%)
Positive Agreement94 .7% Positive Agreement94
Negative Agreement74
.6% Negative Agreement74
95% confidence interval
calculated by the "Score" method.
McNemar Test for Symmetry:
Test < Reference1 (1.3%)
Test > Reference15 (19.2%)
Symmetry test FAILS p < 0.001 (ChiSq=12.250, 1 df)
A value of p<0.05 suggests that one method is
consistently "larger".

Cohen's Kappa55 .5% (36.0 to 74.9%)
Kappa is the portion of agreement above what is
expected by chance The rule of thumb is that Kappa >
75% indicates "high" agreement. We would like to see
VERY high agreement (close to 100%).

	Negative Reference	Positive Reference	Total	Legend		
					Reference	Test
Negative Test	4 4	1	4 5		1 Negative (<=1)	Negative
Positive Test	1 5	1 8	3 3		2 Positive (>1)	(<=23) Positive (>23)
Total	5 9	1 9	7 8			

Number excluded or missing: 0

Experiment Description	
Reference Method	Test Method

Prepared for: Central Laboratory -- Hospital El Carmen
By: Analytical Quality -- Productos Roche S.A.Q.e I

	Experiment Description	
Analyst:	Technician	Technician
Date:	06 Aug 2014	06 Aug 2014
Comment		

Qualitative Method Comparison

Ref Method Microscopy Test Method UF1000

Experimental Results

Spec ID	Ref	Test	Spec ID	Ref	Test	Spec ID	Ref	Test
*0214	0	9,0	*7964	10	247,0	*8533	0	6,1
*1363	0	21,4	*8168	0	3,4	*8534	F0	53,6
*1462	F1	56,4	*8305	1	5,1	*8857	5	27,7
*2017	F0	39,2	*8321	60	281,7	*8967	F1	40,5
*2305	1	10,6	*8325	5	74,1	*8987	10	58,7
*2306	F1	35,5	*8336	5	31,0	*9985	0	5,8
*2327	1	2,7	*8344	1	10,7	0820	F3	6,5
*2329	1	10,8	*8357	0	9,0	1462	6	75
*2330	0	8,1	*8412	1	15,4	1894	1	10,4
*2338	0	9,5	*8482	F1	36,2	2716	5	65
*2362	0	5,1	*8489	F0	35,2	3021	F1	27,6
*2367	0	4,0	*8504	5	83,3	3066	F1	167,2
*2376	1	13,5	*8513	4	690,4	3165	F1	28,12
*2377	0	3,3	*8515	1	21,2	3540	3	29,7
*2812	F1	35,2	*8520	1	7,1	3665	1	19,7
*2860	1	4,0	*8521	1	7,1	3856	1	9,5
*3021	F0	27,6	*8522	0	15,4	4143	1	10,2
*3066	F1	167,2	*8523	0	6,4	4153	1	10,4
*3081	1	3,4	*8524	0	1,8	4202	1	12
*3165	F1	28,8	*8526	0	2,9	4725	24	165
*3339	0	1,0	*8527	10	29,1	4769	1	20
*3341	0	18,9	*8528	0	6,8	4770	0	5
*3344	0	3,2	*8529	0	14,7	4771	1	12
*3345	5	108,1	*8530	0	10,0	4772	4	34
*3481	F1	31,6	*8531	1	6,7	8527	10	29,1
*3741	5	28	*8532	0	1,8	8564	4	83,3

X:Excluded F:No Agreement

Qualitative Method Comparison

Ref Method Microscopy Test Method UF1000

Statistical Summary

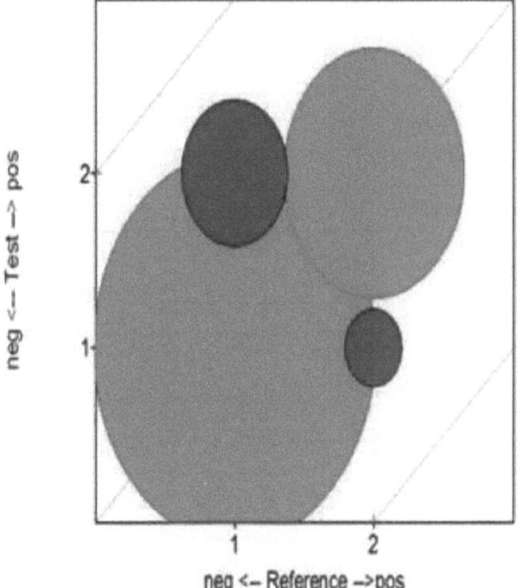

Statistical Analysis
(Comparison of two Laboratory Methods)

Agreement88	.5%	(79.5 to 93.8%)
Positive Agreement90	.9% Positive	
Agreement90	.9% Positive	
Agreement90	.9%	
Negative Agreement87	.5% Negative	
Agreement87	.5% Negative	
Agreement87	.5% Negative	

95% confidence interval calculated by the "Score" method.

McNemar Test for Symmetry:

Test <	Reference2 (2.6%)
Test >	Reference7 (9.0%)

Symmetry test PASSES p = 0.180 (Exact Test)
A value of p<0.05 suggests that one method is consistently "larger".

Cohen's Kappa73   .3% (57.0 to 89.7%)
Kappa is the portion of agreement above what is expected by chance The rule of thumb is that Kappa > 75% indicates "high" agreement. We would like to see VERY high agreement (close to 100%).

	Negative Reference	Positive Reference	Total	Legend	
				Reference	Test
Negative Test	4 9	2	5 1	1 Negative (<=3)	Negative (<=28)
Positive Test	7	2 0	2 7	2 Positive (>3)	Positive (>28)
Total	5 6	2 2	7 8		

Number excluded or missing: 0

Experiment Description		
	Reference Method	Test Method
Analyst:	Technician	Technician
Date:	06 Aug 2014	06 Aug 2014
Comment		

Qualitative Method Comparison

Ref Method Microscopy   **Test Method UF1000**

Experimental Results

Spec ID		Ref	Test	Spec ID		Ref	Test	Spec ID		Ref	Test
*0214		1	0,2	*2824		6	29	*8344		1	2,0
*1363		1	4,2	*2825		1	13	*8357		1	0,4
*1462		30	651,8	*2826		1	5	*8412		40	3629,0
*2017		1	4,2	*2827	F	3	29	*8482		25	205,5
*2305		1	9,9	*2828		1	10	*8489		80	300,5
*2306		1	1,2	*2829		8	36	*8504		1	7,5
*2327		1	4,6	*2830		2	5	*8513	F	3	855,5
*2329	F	1	29,7	*2831		1	17	*8515		2	9,6
*2330		1	0,2	*2832		15	56	*8520		1	5,9
*2338		1	0,1	*2860		1	0,1	*8521	F	4	9,4
*2362		1	0,8	*3021	F	2	33,2	*8522		1	1,7
*2367		1	4,2	*3066		10	540,1	*8523		1	3,3
*2376		4	75,2	*3081		1	1,0	*8524		1	7,2
*2377		1	3,9	*3165		10	140,2	*8526		1	6,9
*2812		2	21,4	*3339		1	0,0	*8527		1	1,0
*2813		2	14	*3341		1	10,7	*8528		1	4,8
*2814		8	34	*3344	F	5	22,8	*8529		1	0,4
*2815		2	12	*3345		5	29,7	*8530	F	2	36,7
*2816		10	47	*3482	F	3	56,2	*8531		40	127,5
*2817		2	21	*3741		1	9,6	*8532		1	2,1
*2818		2	19	*7964		30	597,4	*8533		1	3,2
*2819		25	135	*8168		1	1,8	*8534		1	12,0
*2820		1	10	*8305		1	1,8	*8857		2	4,8
*2821		2	12	*8321		4	106,2	*8967		20	165,4
*2822		4	29	*8325	F	1	29,5	*8987		2	9,9
*2823		5	31	*8336		2	14,4	*9985		2	5,2

X:Excluded F:No Agreement

Prepared for: Laboratorio Central -- Hospital El Carmen By: Calidad Analítica -- Productos Roche S.A.Q.e I

Linearity Summary

Instrument	Analyte	Linearity	Accuracy	Reportable Range	Precision
UF1000	B Bacteria	Pass	Pass	Pass	-
	E Erythrocytes	Pass	Pass	Pass	-

Required parameters are missing. experiment 'fails' (or not enough results).

Experiment 'passes' (or adequate number of results).

Bacteria

Instrument UF1000

EP Evaluator®

	Assigned	Pct	N	Accuracy & Recovery			Linearity	Rpt Range
				Mean	% Rec	Status		
DIL 10	14,77	0,2%	2	14,60	98,9	Pass	Pass	Pass

DIL 9	28,79	0,39%	2	31,20	108,4	Pass	Pass	-
DIL 8	57,59	0,78%	2	60,20	104,5	Pass	Pass	-
DIL 7	115,17	1,56%	2	124,80	108,4	Pass	Pass	-
DIL 6	231,09	3,13%	2	252,50	109,3	Pass	Pass	-
DIL 5	461,44	6,25%	2	443,70	96,2	Pass	Pass	-
DIL 4	922,88	12,5%	2	976,90	105,9	Pass	Pass	-
DIL 3	1845,75	25%	2	1802,50	97,7	Pass	Pass	-
DIL 2	3691,50	50%	2	3540,00	95,9	Pass	Pass	-
DIL 1	7383,00	100%	2	7383,00	100,0	Pass	Pass	Pass

See User's Specifications for Pass/Fail criteria.

Linearity Summary

	Overall
Slope	1,020
Intercept	0,47
Obs Err	6,4%
N	10

LINEAR within Allowable Systematic Error of 10.0%.

User's Specifications

Allowable Total Error	20,0%
Systematic Error Budget	50%
Allowable Systematic Error	10,0%
Reportable Range	300 to 10000 cells
RR-Low Range	0.0 to 600.0 cells
RR-High Range	0,0 to 20000,0 cells

Experimental Results

DIL 10	14,6	14,6
DIL 9	30,2	32,2
DIL 8	60,2	60,2
DIL 7	124,8	124,8
DIL 6	252,5	252,5
DIL 5	473,7	413,7
DIL 4	976,9	976,9
DIL 3	1802,5	1802,5
DIL 2	3540	3540
DIL 1	7383	7383

X: Excluded from calculations

Supporting Data

Analyst	Technician
Date	07 Aug 2014
Value Mode	Pct-Assigned
Units	cells
Controls	-
Reagent	-
Calibrators	-
Comment	

Evaluation of Results

The Accuracy, Reportable Range, and Linearity of Bacteria were analyzed on UF1000 over a measured range of 14.60 to 7383.00

cells. This analysis assumes accurate assigned values. Allowable systematic error (SEa) was 10.0%. The accuracy test PASSED. The maximum deviation for a mean recovery from 100% was 9.3%. 10 of 10 mean recoveries were accurate within the SEa. 20 of 20 results were accurate within the allowable total error (TEa) of 20.0%. The results are LINEAR. The system PASSED reportable range tests.

Evaluation of results

Accuracy, reportable range and linearity of bacteria were analyzed on UF1000 over a measured range of 14.60 to 7383.00 cells. This analysis assumes accurate assigned values. The systematic error allowable (SEa) was 10.0%. The accuracy test PASSED. The maximum deviation for a 100% mean recovery was 9.3%. 10 of 10 mean recoveries were accurate within SEa. 20 of 20 results were accurate within the total allowable error (TEa) of 20.0%. The results are LINEAR. The system PASSED the reportable range tests.

Bacteria
Instrument UF1000
Prepared for: Laboratorio Central -- Hospital El Carmen By: Calidad Analitica -- Productos Roche S.A.Q.e I

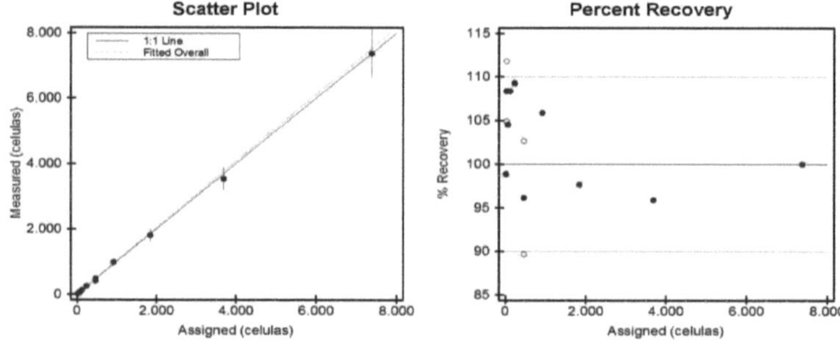

Accuracy, Reportable Range, and Linearity

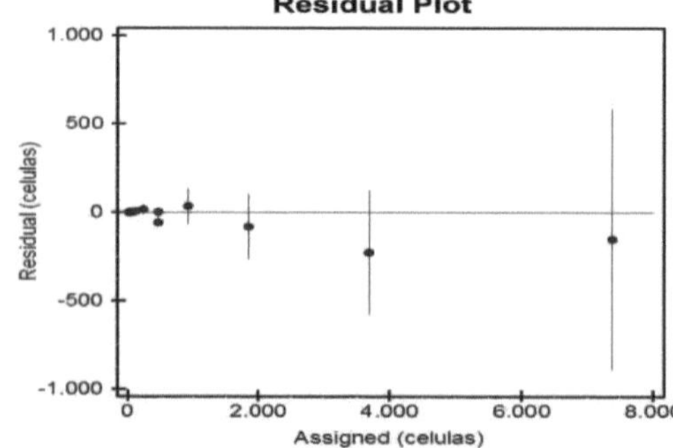

Erythrocytes
Instrument UF1000
EP Evaluator®

	Assigned	Pct	N	Accuracy & Recovery			Linearity	Rpt Range
				Mean	% Rec	Status		
DIL 9	18,656	0,39%	2	19,700	105,6	Pass	Pass	Pass
DIL 8	37,313	0,78%	2	41,300	110,7	Pass	Pass	-
DIL 7	74,626	1,56%	2	83,500	111,9	Pass	Pass	-
DIL 6	149,730	3,13%	2	150,900	100,8	Pass	Pass	-
DIL 5	298,981	6,25%	2	320,200	107,1	Pass	Pass	-
DIL 4	597,962	12,5%	2	621,900	104,0	Pass	Pass	-
DIL 3	1195,925	25%	2	1225,900	102,5	Pass	Pass	-
DIL 2	2391,850	50%	2	2436,800	101,9	Pass	Pass	-
DIL 1	4783,700	100%	2	4783,700	100,0	Pass	Pass	Pass

See User's Specifications for Pass/Fail criteria.

Linearity Summary		Experimental Results		
	Overall	DIL 9	19,90	19,5
Slope	1,049	DIL 8	40,30	42,30
Intercept	1,099	DIL 7	83,50	83,50
Obs Err	4,9%	DIL 6	150,90	150,90
N	9	DIL 5	320,20	320,20
LINEAR within Allowable Systematic Error of 12.5%.		DIL 4	621,90	621,90
		DIL 3	1225,90	1225,90
		DIL 2	2436,80	2436,80
		DIL 1	4783,70	4783,70
				X: Excluded from calculations

User's Specifications

Allowable Total Error	25,0%
Systematic Error Budget	50%
Allowable Systematic Error	12,5%
Reportable Range	0,00 to 4000 cells
RR-Low Range	-20,000 to 20,000 cells
RR-High Range	-783,000 to 8783,000 cells

Supporting Data

Analyst	Technician
Date	07 Aug 2014
Value Mode	Pct-Measured
Units	cells
Controls	-

Reagent	-
Calibrators	-
Comment	

Evaluation of Results

The Accuracy, Reportable Range, and Linearity of Erythrocytes were analyzed on UF1000 over a measured range of 19,700 to 4783,700 cells. This analysis assumes accurate assigned values. Allowable systematic error (SEa) was 12.5%. The accuracy test PASSED. The maximum deviation for a mean recovery from 100% was 11.9%. 9 of 9 mean recoveries were accurate within the SEa. 18 of 18 results were accurate within the allowable total error (TEa) of 25.0%. The results are LINEAR. The system PASSED reportable range tests.

Evaluation of results

The accuracy, reportable range, and linearity of erythrocytes were analyzed on UF1000 over a measured range of 19,700 to 4783,700 cells. This analysis assumes accurate assigned values. The permissible systematic error (SEa) was 12.5%. The precision test PASSED. The maximum deviation for a mean recovery of 100% was 11.9%. 9 of 9 mean recoveries were accurate within SEa. 18 of 18 results were accurate within the total allowable error (TEa) of 25.0%. The results are LINEAR. The system PASSED the reportable range tests.

Prepared for: Laboratorio Central -- Hospital El Carmen By: Calidad Analítica -- Productos Roche S.A.Q.e I

Accuracy, Reportable Range, and Linearity

Erythrocytes Instrument UF1000

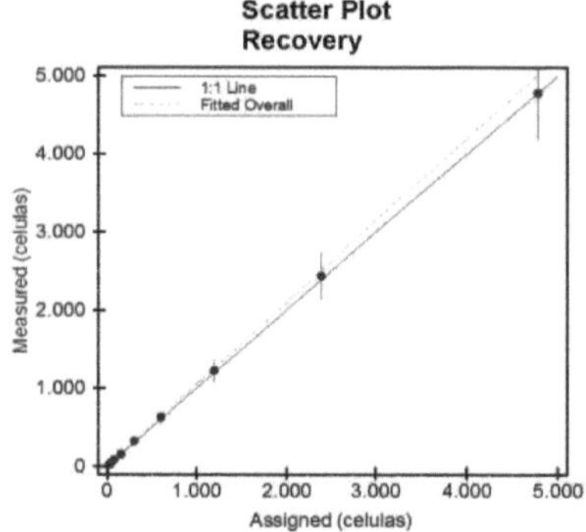

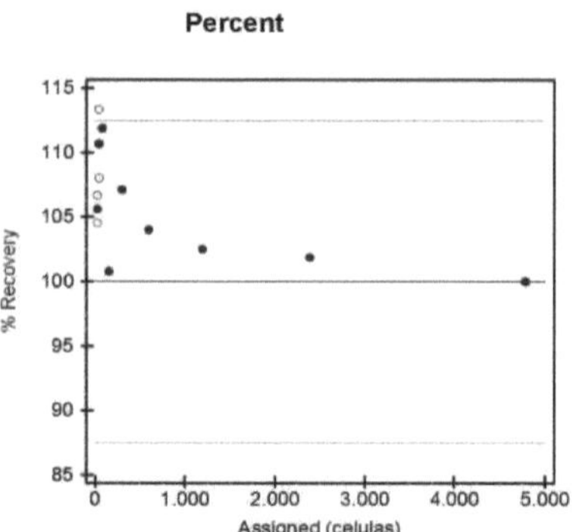

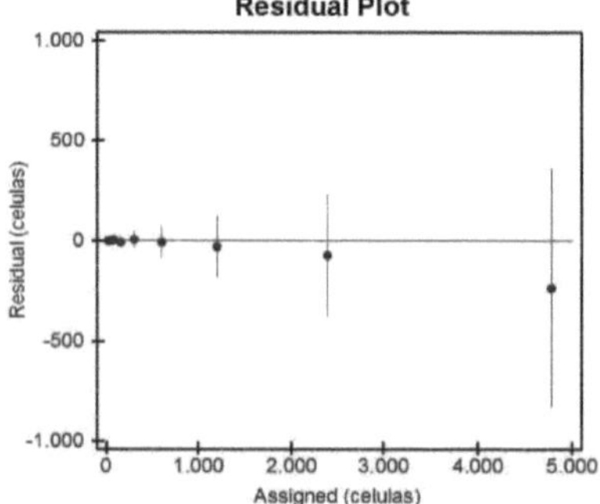

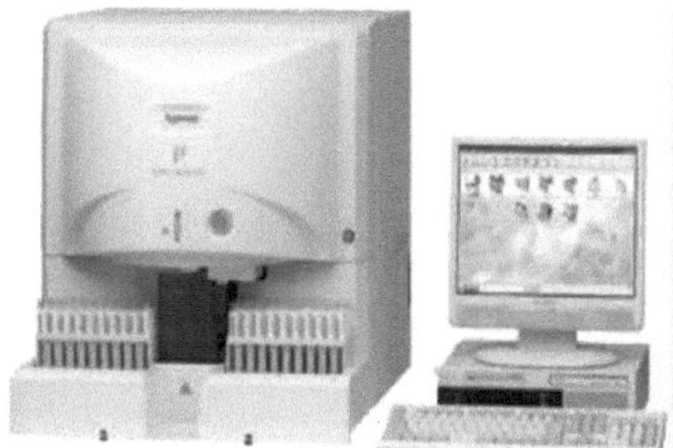 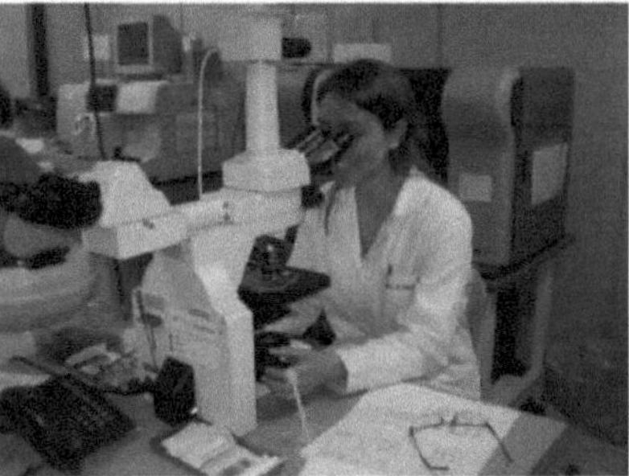

CONCLUSIONS

The adaptation of the UF1000 equipment was successful. In the chosen conditions it presents acceptable linearity and accuracy, providing comparable results obtained by optical microscopy.

PERSONAL EXPERIENCE

Regarding the parameters "hyaline cylinders" and "pathological cylinders" we observed that the limit obtained by statistics was not acceptable for our population since this parameter has clinical significance orienting the medical diagnosis, so it was decided to reduce the lower limit to value "0", this means that any measurement of the equipment referred to these parameters should be reported as a microscopic observation alarm.

BIBLIOGRAPHY .

https://medlineplus.gov/spanish/urinalysis.html

https://www.farestaie.com/cd-interpretacion/te/bc/295.htm

IMAGES

- MEASUREMENT OF PROTEINURIA IN THE DIAGNOSIS OF PREECLAMPSIA, GUIDE ...

encolombia.com
- Urine Epithelial Cells - Epithelial Cells - celulasepiteliales.com
- ANOMALIES IN URINE - Urinary Sediment Atlas sites.google.com
- The norm of leukocytes in the urine of children. Leukocytes in ... fehrplay.com
- Atlas Urinary Sediment. en.slideshare.net
- Clinical Assessment: 2014- valoracionclinicaeq2.blogspot.com
- Urine | VETERINARY ANALYSIS LABORATORY lav-asoria.com
- Urinary sediment. Dr. Chaverri.
medicine-ucr.com
- Urinalysis - Chemistry and something more quimicayalgomas.com
- Test farestaie.com
- URINE LEUKOCYTE CASTS PDF
nikefreerun3runningshoes.xyz
- Test farestaie.com
- Urine crystals hippuric acid - Sociedad Argentina de Citología sociedaddecitologia.org.ar
- Amorphous phosphates in children's urine: causes and what it means medsaludin.es
- phosphatetriple - Hash Tags - Deskgram deskgram.net
- ATLAS OF CRYSTALS AND URINARY CALCULATIONS qualitat.cc
- biurato hashtag | Picgra picgra picgra.com
- Test farestaie.com
- Clinical case cystinuria | Belladonna Lilly
belladonnalilly.wordpress.com
- General Urine Test Interpretation slideshare.net
- CHOLESTEROL CRISTS (Acid Urine Crystitis) - Sediment ...

pinterest.com.mx
- Bacteria in urine - YouTube youtube.com
- Yeast in Urine, Meaning - Candida albicans segundomedico.com
- Mucin Filaments abundant in Urine - SegundoMédico segundomedico.com
- spermatozoa in urine - SegundoMédico segundomedico.com

Buy your books fast and straightforward online - at one of world's fastest growing online book stores! Environmentally sound due to Print-on-Demand technologies.

Buy your books online at
www.morebooks.shop

Kaufen Sie Ihre Bücher schnell und unkompliziert online – auf einer der am schnellsten wachsenden Buchhandelsplattformen weltweit! Dank Print-On-Demand umwelt- und ressourcenschonend produzi ert.

Bücher schneller online kaufen
www.morebooks.shop

Printed by Books on Demand GmbH, Norderstedt / Germany